For Georgie, Emilio and Iona. G.C.

For Lottie - here's to building our own home. Thank you for your support, kindness and patience, always. R.S.H.

First published in Great Britain 2024 by Red Shed, part of Farshore
An imprint of HarperCollins*Publishers*
1 London Bridge Street, London SE1 9GF
www.farshore.co.uk

HarperCollins*Publishers*
Macken House, 39/40 Mayor Street Upper, Dublin 1, D01 C9W8

Text copyright © George Clarke 2024
George Clarke has asserted his moral rights.
Illustrations © HarperCollins*Publishers* 2024
Illustrated by Robert Sae-Heng.

ISBN 978 0 00 858789 5
Printed in Italy.
001

A CIP catalogue record for this title is available from the British Library.

All rights reserved. No part of this publication may be reproduced, stored in a retrieval system, or transmitted, in any form or by any means, electronic, mechanical, photocopying, recording or otherwise, without the prior permission of the publisher and copyright owner.

Stay safe online. Any website addresses listed in this book are correct at the time of going to print. However, Farshore is not responsible for content hosted by third parties. Please be aware that online content can be subject to change and websites can contain content that is unsuitable for children. We advise that all children are supervised when using the internet.

Adult supervision is advised for activities within the book. Always ask an adult for help when using glue, paint and scissors. Wear protective clothing and cover surfaces to avoid staining.

This book is produced from independently certified FSC™ paper to ensure responsible forest management.

For more information visit: www.harpercollins.co.uk/green

GEORGE CLARKE

HOW TO BUILD A HOME

ILLUSTRATED BY
ROBERT SAE-HENG

INTRODUCTION

When I was young, I would often hear the words 'house' and 'home'. If I'm honest, at the time I thought they were the same thing. The end of the day was called 'home time', but then my friends would sometimes say, "George, let's go to your house after school." Did I live in a 'house' or a 'home'?

It wasn't until I was 12 years old, when my grandad bought me a book about architecture, that I discovered the difference. A house describes the physical structure of a building someone lives in. A house becomes a home when you move in, and it becomes a special place to you.

Growing up, my bedroom was my world. It was my private space, full of things I loved and cherished, and somewhere I was given permission to decorate and change the layout of. I adored the process of making my bedroom feel different through design.

I would measure my room and then draw a plan to show where the door and windows were. I would then sketch out lots of different positions for my bed, my wardrobe and even my drum kit. It was fun seeing how my room

could be laid out differently to change the way it would feel, and it was through this process of redesigning my bedroom that I knew I wanted to be an architect.

An architect is a person who is trained to design buildings, and those buildings could be anything from a skyscraper or an airport to a museum and even a factory. When I became an architect, I realised that I loved designing homes. For me, a home is the most important piece of architecture, because whether it is big or small, new or an adaptation of an old home, even the smallest change can make a big impact. Imagine if you moved a door or window in your home, for example. How different would it make your home look and feel?

I hope while you're reading the pages ahead that you might be inspired. If you are lucky enough to have your own bedroom, then you could imagine ways of transforming it, or perhaps think about how you might change your home or someone else's. Perhaps you could even imagine what your dream home of the future could look like? What materials would you use? What shape would it be?

Together, we're going to look at the homes around us and begin to understand the stories behind them.

INCREDIBLE HOMES

Homes can be designed and built in all different shapes, sizes and colours. They can be square, long, round ... anything! This is what makes architecture so exciting.

Homes have always been designed according to what materials and technology are available, and the climate.

Some of the earliest human homes were tents covered in animal hides, which could be easily moved around, and circular wooden huts plastered with mud.

People have lived in treehouses for thousands of years. The first ones kept people safe from dangerous animals or flooding.

I designed a tiny home that looked futuristic. It was a cylinder, with a three-metre cube inside that rotated to create four different rooms in one space. The black slab changed from a table to a TV to a bed!

In hot countries, homes may have small windows and thick walls to stop too much heat coming in, and external walls may be painted white to reflect the sun and help keep homes cool inside.

There are lots of white homes in hot countries such as Greece and Spain.

In places that have lots of snow, homes often have pointy roofs so snow doesn't build up and get too heavy.

In 17th century Norway, white or yellow paint was used to show wealth, as it could be seen from far away and gave the impression the home was made from expensive stone.

And many homes were red because it was the cheapest paint to make. In fishing areas, ochre was mixed with cod liver oil.

Experimenting with the size of homes led architects to create 'tiny homes', which are popular as they are easier and quicker to build, use less materials, and need less energy to heat, cool and power. This makes them more affordable and sustainable.

American inventor Elisha Otis invented the first safety lift in the 1850s.

Thanks to the invention of the lift and crane, and the use of steel, there has been a boom in high-rise homes since the late 1800s.

Most homes over one storey have the living spaces downstairs, but I've seen 'upside-down' homes, where the living spaces are upstairs to make the most of a good view.

Next time you walk around your local area or visit somewhere new, see how many types of homes you can spot.

If you live in a town or city, you might live in, or see, a high-rise building of flats. The first-known high-rise homes were in ancient Rome. Many were around five storeys but some were up to nine storeys. The safety lift hadn't been invented yet, so they had to go up a lot of stairs!

THE SKY'S THE LIMIT

What sort of home do you live in? Is it in the countryside, a village, a town or a busy city? Is it on one level, like these flats, or over more than one storey?

TEAMWORK

When I design a new home for a client, I start by putting my ideas down on paper; drawing plans. I constantly re-sketch and test my ideas to make the design work better. This isn't fast, but it is fun!

Imagine, or draw, your dream home. No matter how adventurous it is, there will be a way to build it.

Once I'm happy with my plans, I then work with some incredible people to make my design come to life . . .

Heating, ventilation and electrical systems are designed by a mechanical and electrical engineer.

An ecological consultant can help make the build as sustainable as possible.

A quantity surveyor works out how much it will cost to build.

A structural engineer helps make sure the structure is safe to build.

MARVELLOUS MATERIALS

There are so many materials I can choose from when designing a home. What materials are in your home? Perhaps glass in the windows and bricks on the outside walls? Anything else?

Historically, people built using natural materials that were nearby, such as timber, mud and clay. It was an ecological way to build because the materials were in abundance and didn't have to be transported far.

Timber comes from trees and must be dried out before it can be used, otherwise it shrinks too much. It can last a long time but needs protecting from fire and wood-eating insects.

Mud bricks are a mixture of mud (clay, sand and water) and a binding material, such as straw. The earliest-known mud bricks are from 9000 BCE near Jericho, Palestine. People stomped on the mixture, then shaped it into bricks that dried in the sun.

When people started to fire mud bricks in a kiln about 3500 BCE, it made the bricks stronger. It also meant the bricks could be made in cooler climates, where it wasn't possible to use the sun's heat to dry them.

One mud brick could support the weight of one elephant; one fired mud brick the same size could support the weight of five elephants.

Clay is a soft material that comes from the ground. It also becomes hard when dried out or fired in a kiln. Clay has been used for thousands of years to make bricks, and it is one of the main materials still used for bricks today. Bricks are now usually made in factories.

Fired clay bricks are still a popular building choice, as they aren't too expensive to make and last a long time; this is why we see many old and new brick homes today.

Glass revolutionised home building, as it lets light in and allows us to see out, while at the same time making a home wind and watertight.

The first people we know of that used glass in windows were the ancient Romans (around 100CE).

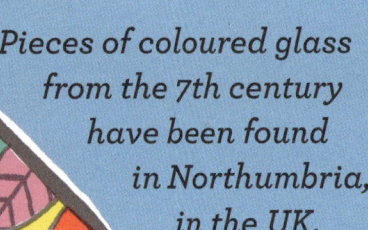

Pieces of coloured glass from the 7th century have been found in Northumbria, in the UK.

Glass is made from heating up raw materials (sand, limestone and sodium carbonate) at a super high temperature in a furnace until they melt into a liquid. This then floats on top of liquid metal to keep it flat. Once the glass has cooled slowly, it is cut into different shapes.

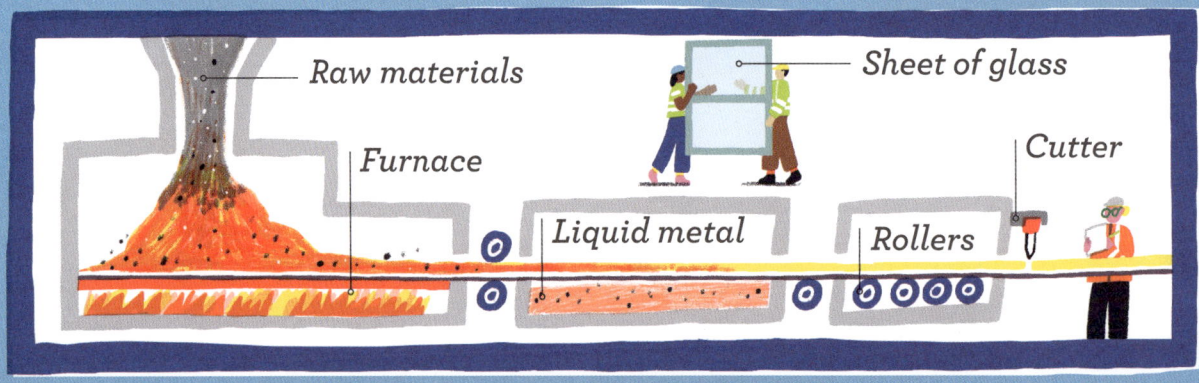

Concrete and steel became popular materials in the 20th century because they were cheap and strong. Their strength meant that homes could be built with fewer structural support walls, and more glass could be used. This completely transformed the design of homes.

Concrete is a mix of Portland cement, aggregates, such as crushed stone, gravel and sand, and water. When concrete is first mixed, it is a wet, paste-like material, but as it dries, the cement binds to the aggregates and becomes solid like rock.

Concrete is usually made in a factory and transported to building sites in a concrete-mixing lorry.

Making concrete needs a huge amount of energy, so it isn't a very ecological building material. The energy needed to make a one-metre cube of concrete is about the same as a lightbulb turned on all day and night for nearly three months!

Wait at least 24 hours for fresh concrete to set before walking on it!

THINK LIKE AN ARCHITECT

Imagine building a home on this hillside for your family, where the weather can be warm, wet or freezing cold, depending on the time of year.

What size and shape would it be?

What materials would you use?

Where would you add windows to make the most of the views and natural light?

How many rooms would you have?

Would any part of your home be underground?

Grab some paper and a pencil, and start designing. I love making models to test ideas too. Let your imagination run wild!

When I'm designing a home, I spend time at the place where it will be built, to help form a picture in my head of what it might look like.

Architects also think about the layout of rooms and where things should be positioned, such as furniture and lights.

Have a go at re-designing a room in your home. You'll need: a measuring tape, plain paper, grid paper and a pencil.

Step 1: Measure the room; write down its length and width on the plain paper.

Step 2: On the grid paper, draw your room to scale. 1 square = 10 centimetres. So, if your room is 200 centimetres wide, it would be 20 squares wide (1:10 scale).

Step 3: Measure the different pieces of furniture and draw them to scale on the grid paper.

Step 4: Now try out different designs with the furniture in other positions. You'll see that some layouts work better than others.

Scale is important. If we drew a home at 1:1 scale, the drawing would be life size!

Ask an adult to help.

PRINT YOUR HOME

Any object can be printed these days – even entire homes! Imagine using a 3D printer to bring your design to life.

First of all, you would need to create your design on a computer: from scratch using computer-aided design (CAD) software or re-creating a hand-made drawing.

The 3D printer would then create the design in 3D by printing layer after layer of material until it's fully formed.

A 3D printer is like a robot builder, but the design still has to be done by a person.

Small 3D printer

Small 3D printers are perfect for printing objects such as door handles and light switches.

Think of a 3D printer like squeezing toothpaste onto a toothbrush, and then adding more and more layers of toothpaste on top of each other. A huge 3D printer is needed to make a home!

Most 3D printers use concrete to print homes (which isn't good for the environment).

It's possible to print homes using eco-friendly materials – some companies have used mud or recycled plastic bottles.

As well as printing full-size homes, a 3D printer can make small models to test designs.

Although 3D printing is already being used, we have only just started to explore its potential. The future of 3D printing is exciting because it can make amazing shapes quickly.

ECO HOMES

As well as trying to use sustainable materials to build homes, we need to think about other ways to make them more eco-friendly.

The Sun sends enough energy to Earth in one hour to power the electricity needs of every person in the world for one year. That's an amazing gift, and solar panels can use some of this energy to give us electricity, but we don't yet know how to make the most of it.

Most of a home's energy usage goes towards heating, but it is often lost through tiny gaps in the roof, doors, windows and floors.

To reduce heat loss, homes need insulation; foam panels or spray foam are often used, but there are eco options – like jeans! Denim insulation is created with recycled material left over from making denim clothes.

Why waste rainwater? It can be collected from the roof, stored in a tank in the home and used for washing machines, flushing toilets and watering the garden.

Rainwater can also be collected in water butts to use in the garden.

There are special, super energy-efficient types of home that can stay at a constant temperature all year and need little additional heating or cooling.

Have you seen solar panels on a roof? These are sheets covered in cells that convert light from the sun into electricity to power appliances, lights and and devices.

Some new homes don't need these separate solar panels, as they have roof tiles with solar cells on them.

My home generates heat and hot water using an air source heat pump. It captures heat from the outside air to warm up water in a cylinder for my radiators, bath and shower.

An air source heat pump works even when it is chilly outside – isn't that amazing?

Hot water cylinder

Air source heat pump

FUTURE MATERIALS

Development of new materials will change the future of home building. I love the idea that nature and biology can inspire our homes to be better for the environment.

Mycelium bricks

Imagine using mushrooms as a building material! Mycelium is part of a fungus and this rootlike, fibrous material looks similar to white mould in its raw state.

Amazingly, it can be combined with other materials, such as corn husks, to form building blocks that are strong and compostable.

Hy-Fi

Dried mycelium is very resistant to water, mould and fire, making it a great building material.

A team in New York once used 10,000 mycelium and corn husk bricks to create a temporary, 13-metre tower, called Hy-Fi, which was designed by architect David Benjamin. When the tower was taken down, the bricks were turned into compost, which was then used in community gardens. How cool is that? Wouldn't it be great if mycelium bricks were used to build homes in the future?

Or how about using chocolate for building? Well, shells of cacao tree seeds (which are grown to make chocolate) to be exact! In Ecuador, South America, there are plans to use cocoa bean shell waste to 3D print a village.

Or a super material inspired by a pencil? Let me explain . . . Like graphite inside a pencil, graphene is made from carbon. Graphene could be used to build homes that are stronger, lighter, need less material and last longer.

Graphene is 200 times stronger and six times lighter than steel.

Graphene has only recently been discovered and it is super expensive to make. Hopefully, with more research, it will become affordable.

FUTURE TECHNOLOGY

Home technology is moving forward fast. We can already switch on heating with a phone or turn on lights just by walking into a room. As technology develops, the way we live will change – hopefully it will make life easier, healthier and more comfortable.

Perhaps when you go into a room, technology will track your health and sense your mood? Maybe it would then change lighting colours or release smells to help you feel better. Wouldn't that be incredible?

Perhaps scanners will measure body temperature and adjust the heat of a room automatically.

Detecting diseases early gives us a chance to live longer. Doctors can analyse wee and poo for signs of illness, but a company in Japan has developed a 'wellness toilet' that can do this. Wouldn't it be brilliant if every home had one in the future?

Future technology may allow us to create homes in space. As well as designing beautiful and sustainable homes on Earth, we may one day live on Mars or the Moon.

What technology do you think might be useful in homes in the future?

Companies are already working on designs to build homes in space.

Whether homes are on Mars or Earth, are round or square, low-rise or high-rise, they should all be environmentally friendly and designed in a way that improves the quality of our lives, for now and for many generations to come.

GLOSSARY

3D – We live in spaces that have three dimensions or measurements. These are width, length and depth.

aggregates – Material formed from many small fragments, such as broken rock, sand or gravel.

BCE/CE – BCE means Before the Common Era (before year 1 of our calendar) and CE means after the Common Era.

ecological (eco) – The relationship between living things and their environment.

fungus (plural fungi) – An organism, such as mushrooms, mould and yeast, that gets its food from decaying material or other living things.

insulation – A material used to keep something warm.

Portland cement – The most common type of cement (a powder that hardens when you add water).

safety lifts – Lifts with a device to stop them falling if the lifting chain or rope breaks.

scale – The size of something. A scale drawing makes something either bigger or smaller than its original size.

solar cells – A device that converts sunlight into electricity.

sustainable – Something that is not likely to harm the environment.

tiny homes – Very small buildings that people can live in. They are often less than 40 square metres in floor area, although there is no exact definition.

About the Author
George Clarke is an architect, writer, lecturer and TV presenter. George wanted to be an architect from the age of 12, and after studying architecture at university, he started his own practice.

About the Illustrator
Robert Sae-Heng is a Thai–Mexican illustrator currently based in the UK, who began using imagery to communicate as a child because he spent his early years unable to hear. Robert combines hand-drawn elements with a digital finish.